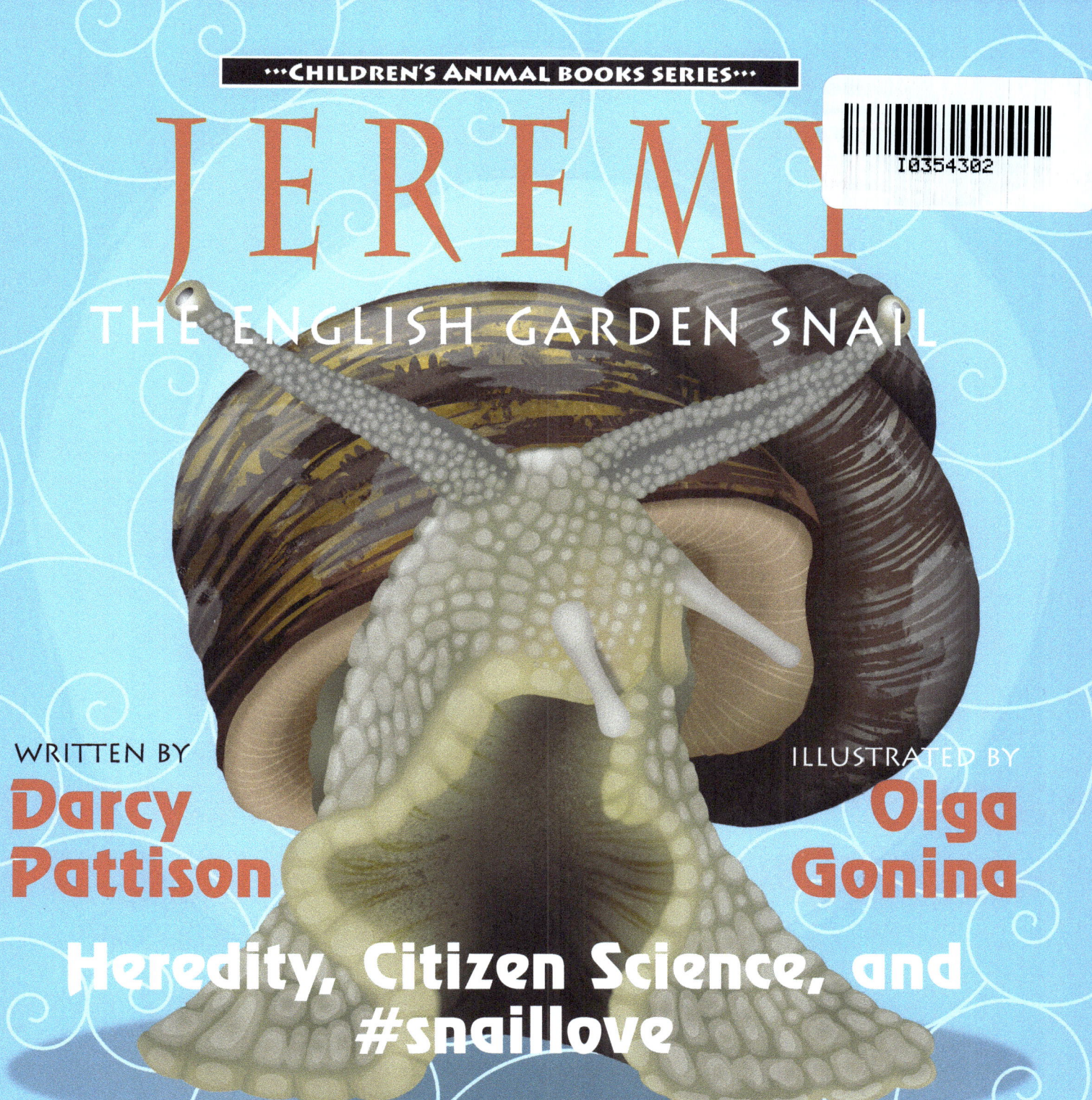

Jeremy, the English Garden Snail: Heredity, Citizen Science, and #snaillove

Text copyright © 2025 by Darcy Pattison
Illustrations copyright © 2025 by Mims House
All rights reserved.

Mims House Books
1309 Broadway
Little Rock, AR 72202
MimsHouseBooks.com

Publisher's Cataloging-in-Publication Data

Names: Pattison, Darcy, author. | Gonina, Olga, illustrator.
Title: Jeremy the English garden snail : heredity , citizen science and #snaillove / By Darcy Pattison; illustrated by Olga Gonina.
Description: Little Rock, AR: Mims House, 2024. | Summary: When a scientist discovers a rare left-coiled snail, he asks social media for help—#snaillove—in locating another left-coiled snail This is a story about heredity, inheritance, community science, and of course, #snaillove.
Identifiers: LCCN: 2024921733 | ISBN: 9781629442556 (hardcover) | 9781629442563 (paperback) | 9781629442570 (ebook) | 9781629442587 (audiobook)
Subjects: LCSH Snails--Juvenile literature. | Heredity--Juvenile literature. | Genetics--Juvenile literature. | Science--Social aspects--Juvenile literature. | BISAC JUVENILE NONFICTION / Animals / General | JUVENILE NONFICTION / Science & Nature / Zoology | JUVENILE NONFICTION / Science & Nature / Biology | JUVENILE NONFICTION / Science & Nature / Discoveries
Classification: LCC QL430.4 .P56 2024 | DDC 594/.3--dc23

**SOURCES**

Davison, Angus, Philippe Thomas, and "Jeremy the Snail" Citizen Scientists. "Internet 'Shellebrity' Reflects on Origin of Rare Mirror-Image Snails." Biology Letters 16, no. 6 (June 2020), https://doi.org/10.1098/rsbl.2020.0110.

Email correspondence and personal interview, December 15, 2023, with Angus Davison, PhD.

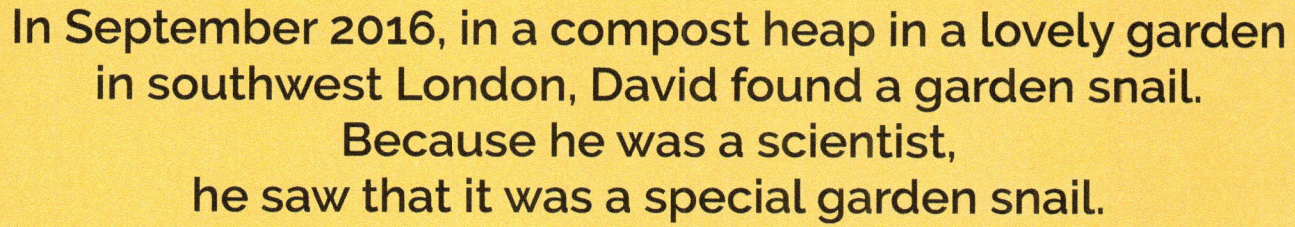

In September 2016, in a compost heap in a lovely garden in southwest London, David found a garden snail. Because he was a scientist, he saw that it was a special garden snail.

Do you see what's special?

Most garden snails have a right-coiling shell.

This garden snail had a left-coiling shell.

The left-coiled snail had situs inversus (SĪ tus In VERS us), which means its body was the opposite of other garden snails—a mirror image.

A child's genes result from combining information from their mother and father. Genes are a set of rules that tell our bodies how to grow and work.
For example, human genes might help decide
if a child will have blue or brown eyes,
become tall or short,
or have curly or straight hair.

For a while, Angus just showed the garden snail to friends. Eventually, he started asking questions.

Did the snail's parents have right- or left-coiled shells?

**There was no way to know.**

If this snail had babies, would they have right- or left-coiled shells?

That was a question Angus could solve.

# Except...

# ...there was one big problem.

Snail shells are asymmetrical, which means they are different on the right and left sides.

Humans are symmetrical on the outside, but on the inside, they are asymmetrical, too. For example, the human heart is to the left side.

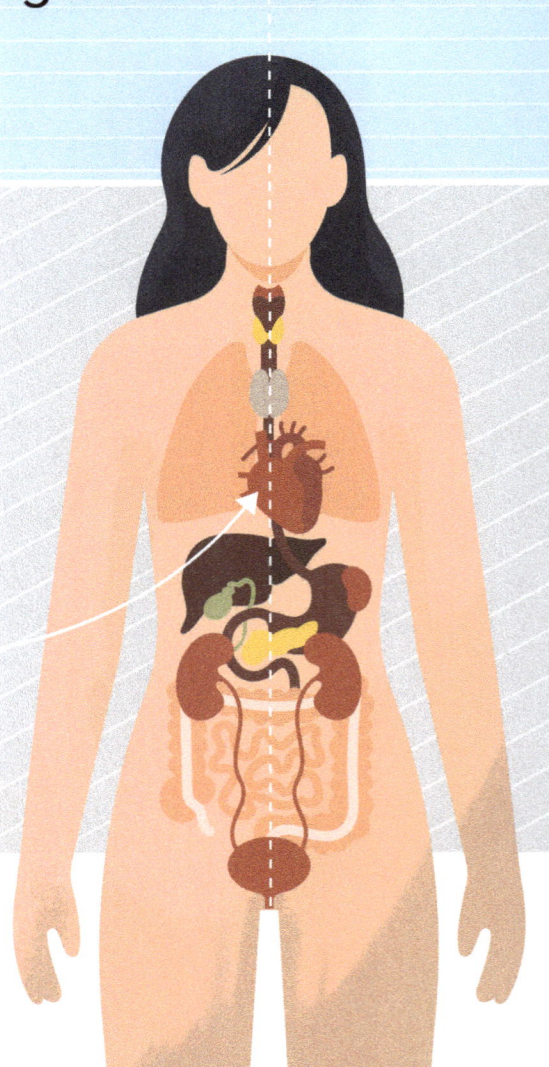

Research on asymmetrical snails could help us understand human bodies.

Because snails are asymmetrical,
left-coiled snails can only mate with left-coiled snails.

The left-coiled snail's shell and body
wouldn't match up with a right-coiled snail.

Angus had worked with snails for over 20 years, though. He had never seen a left-coiled garden snail before.
Angus needed help!

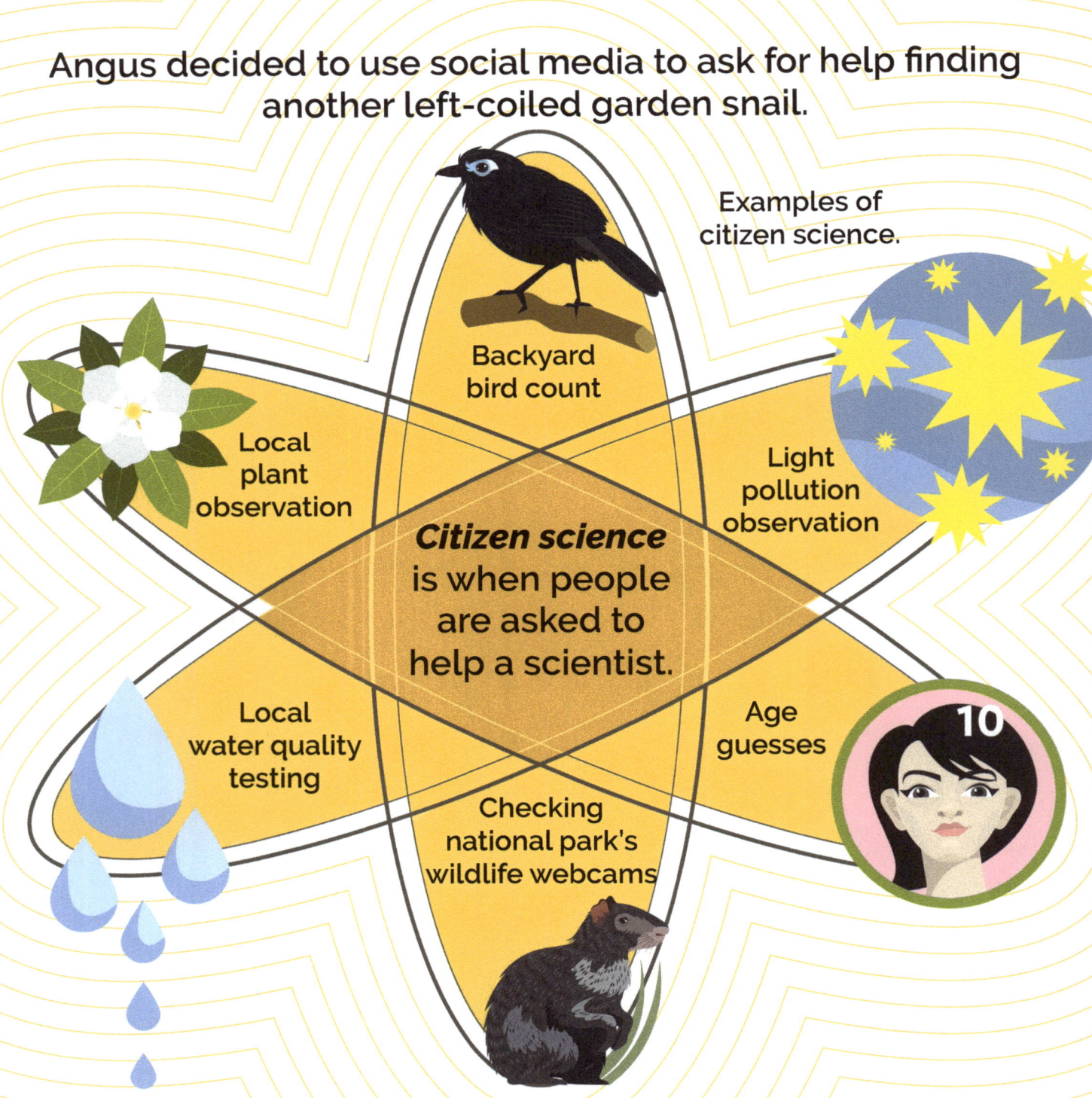

Citizen science works better if the project is explained clearly. Angus and the university staff planned how to explain what they needed to the public.

First, they gave the garden snail a name: Jeremy.

It was named after Jeremy Corbyn, a British politician who loved to garden and had what we call left-wing politics.

Next, Angus created a social media campaign, called #snaillove.

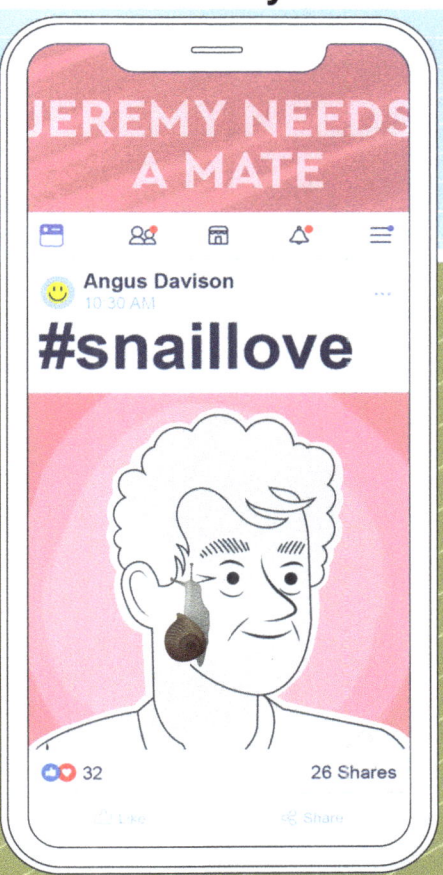

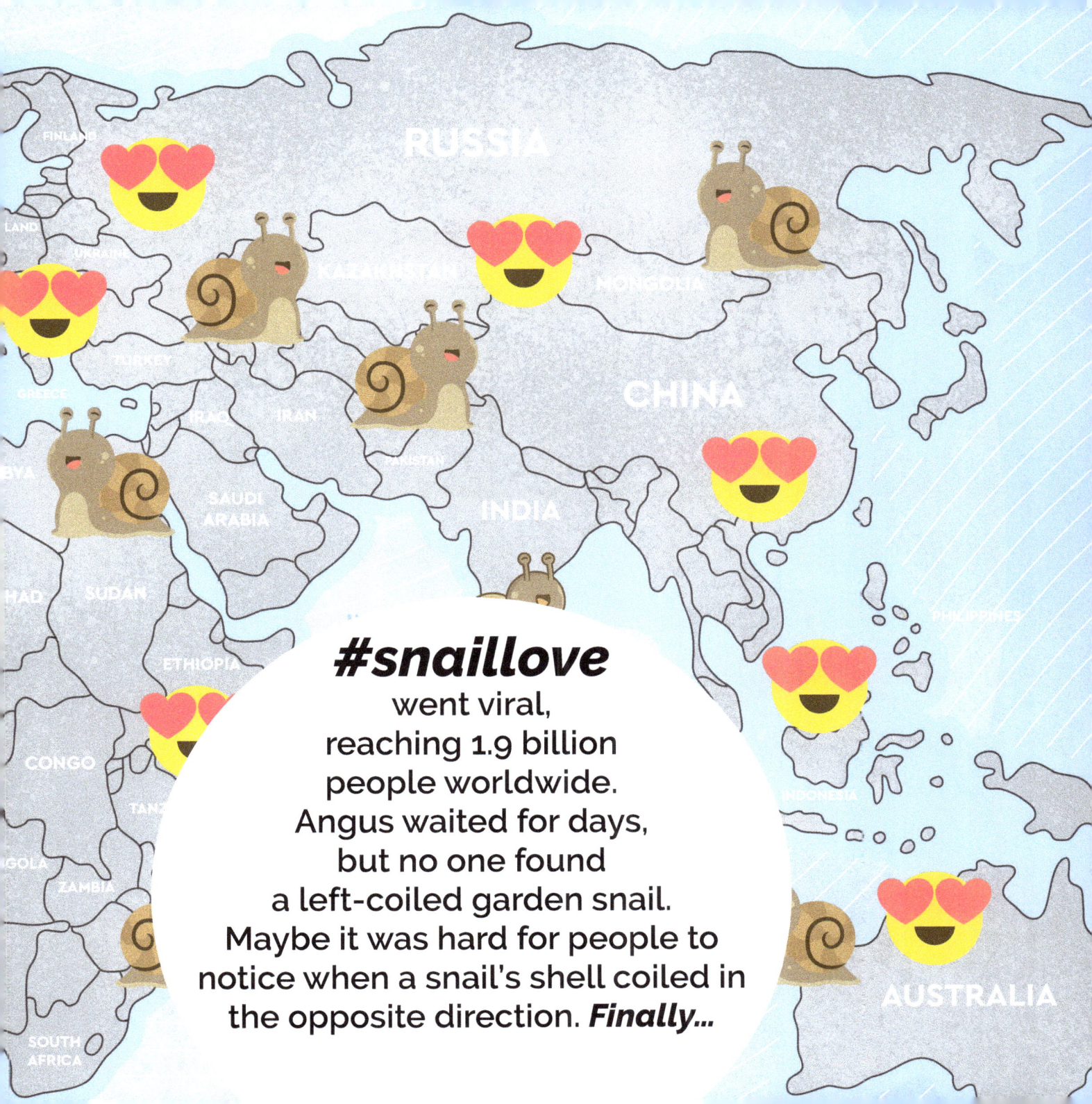

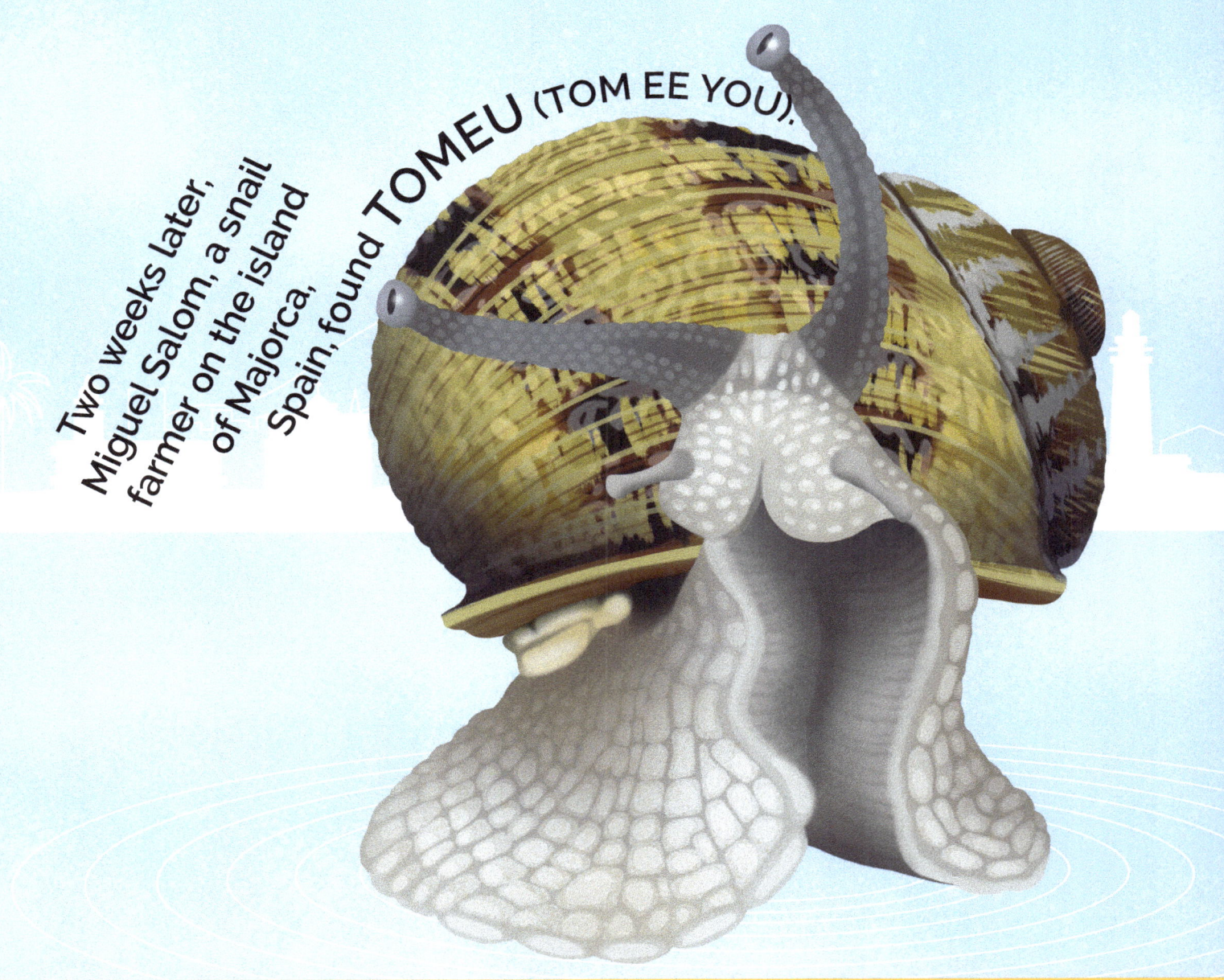

Two weeks later, Miguel Salom, a snail farmer on the island of Majorca, Spain, found TOMEU (TOM EE YOU).

The two left-coiled garden snails were sent to Angus's laboratory by snail mail.
At last, they might discover if the left-coiled shell could be inherited or passed down from a parent.

Land snails are hermaphrodites,
which means they have both male and female parts.
When they mate, both snails can lay eggs.
Lefty and Tomeu mated and had 325 babies.

But Jeremy was left out.

Snails are more likely to mate in the spring when they wake up from hibernation. Angus put Jeremy into a refrigerator to encourage the snail to hibernate.

Waking up three months later, Jeremy met with Lefty and Tomeu again.

But Jeremy still didn't mate.

Then Lefty went home to Ipswich.

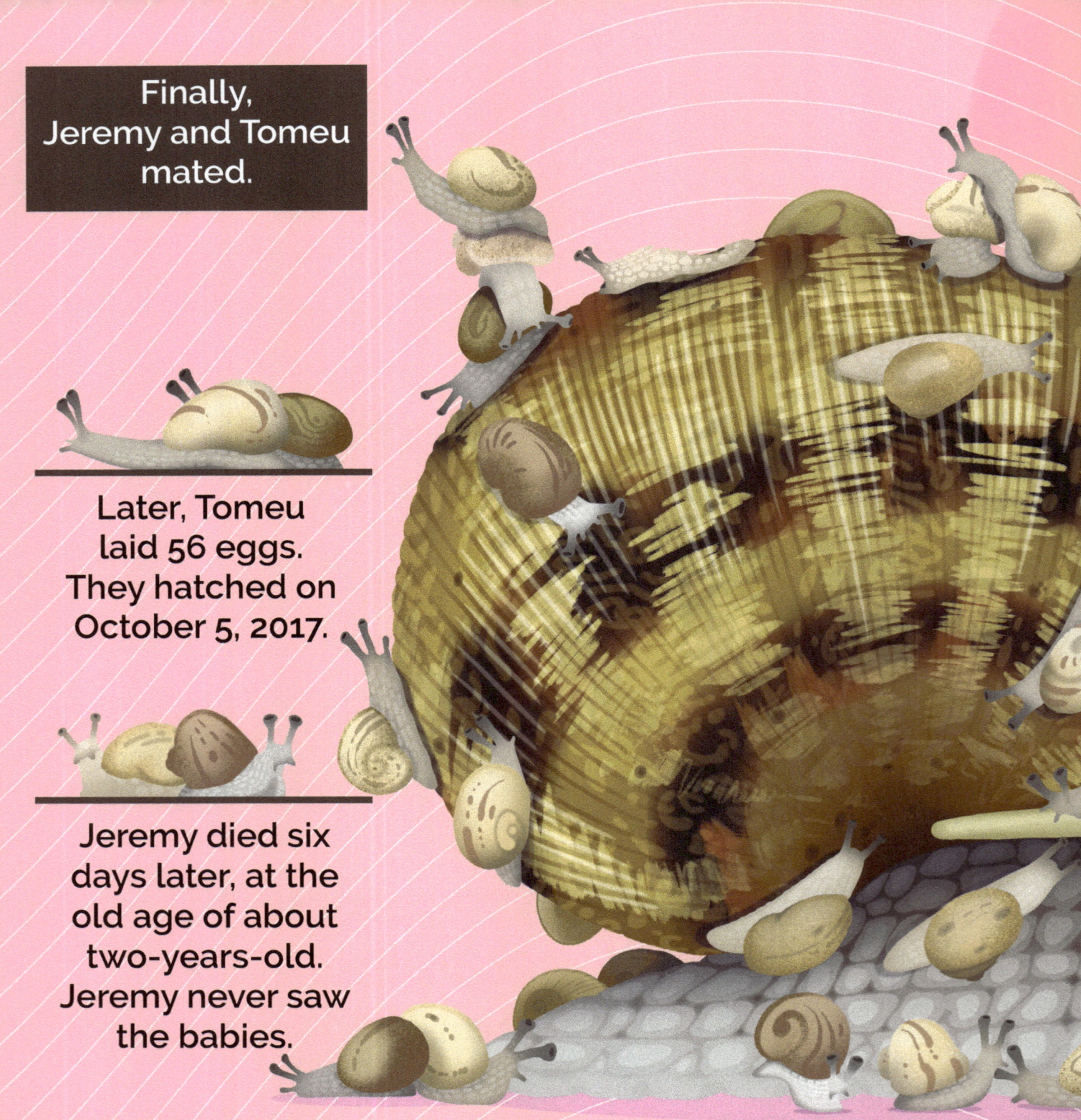

Finally, Jeremy and Tomeu mated.

Later, Tomeu laid 56 eggs. They hatched on October 5, 2017.

Jeremy died six days later, at the old age of about two-years-old. Jeremy never saw the babies.

The three left-coiled garden snails—Jeremy, Tomeu, and Lefty—had 381 baby snails.

All the babies had right-coiled shells.

The genes from Jeremy, the left-coiled garden snail, didn't produce any other left-coiled snails.

But the project was a success because geneticists learned more about asymmetry and how it is inherited.

And worldwide, Jeremy inspired people...

...TO L♥VE SNAILS.

#snaillove
#lovesnails

# GARDEN SNAIL

*Cornu aspersum,* or European brown snail

Garden snails belong to the mollusk phylum, a large group of invertebrate animals that also includes clams, oysters, and octopuses.

### DISTRIBUTION
The garden snail is native to the Mediterranean Sea area, northwest Africa, Asia Minor, western Europe, and Great Britain. In the U.S. and Canada, it is an invasive species.

### LIFESPAN 2-3 years

### HERBIVORES
Garden snails are herbivores, eating fruit trees, vegetables, and flowers.

### HERMAPHRODITES
Most land snails are hermaphrodites, which means they have both male and female parts.

### HIBERNATION
Land snails will hibernate in the winter and may also aestivate during hot weather.

Shell: 1 to 1.6 inches (25 to 40 millimeters) in diameter

When the snail is threatened or inactive, it can pull its entire body into the shell.

The shells has four or five spirals, usually coiling to the right.

Slime trail.

- Eye-like light sensors.
- Feeling and smelling tentacles.
- Soft, slimy body. Brownish gray color

1 to 1.4 inches (25 to 35 millimeters) high

**ASYMMETRY**
Snails are the only animal that has natural variation in asymmetry.

**SNAILS AS FOOD**
Some people cook and eat snails.

**INHERITED? IT DEPENDS**
Science rarely gives a yes-or-no answer. Instead, it depends. The garden snails in the research studies in England did not have a left-coil gene. But the garden snails in similar research studies in France did have a left-coil gene. The *Euhadra* group of Japanese snails regularly produce right- and left-coiled snails; therefore, they have a gene that determines the direction of the shell's coil. In other words, the way a snail's shell coils depends on the type of snail.

Scientists always talk about the research details because they matter. Results always depend on those details.

**CITIZEN SCIENCE**

Citizen science is when the public helps with scientific research. Scientists may ask people to help collect specimens or data, or to help analyze information. One of the longest-running citizen science programs is the Audubon Society's annual U.S. Christmas Bird Count. (https://www.audubon.org/conservation/science/christmas-bird-count) Launched by bird magazine editor Frank Chapman in 1900, it asks people to count birds during specific days close to Christmas and report how many of each species were seen. The Bird Count now includes Canada, Latin America, the Caribbean, and Pacific Islands. For more information on citizen science projects, see Zooniverse.org.

# GLOSSARY

**Asymmetrical:** Not having the same shape, size, or arrangement on both sides.

**Genes:** A set of physical rules that tell our bodies how to grow and work. The physical rules are encoded by DNA within a cell.

**Geneticist:** A scientist who studies heredity and inheritance.

**Heredity:** How traits are passed from parents to children.

**Inheritance:** Receiving traits from parents or other ancestors.

**Situs Inversus (SĪ tus In VERS us):** Body parts are reversed, like a mirror image.

**Social media campaign:** Using social media to share a specific message.

**Snail mail:** This is a slang term for mail that is sent by regular postal services. Because snails have a reputation for moving slowly, it means the slow way to send mail. The term was created to contrast with the speed of email.

### ···CHILDREN'S ANIMAL BOOKS SERIES···

**BIRD** — WISDOM, THE MIDWAY ALBATROSS — Darcy Pattison, illustrated by Kitty Harvill — Surviving the Japanese Tsunami and other Disasters for over 60 Years

**MAMMAL** — ABAYOMI, THE BRAZILIAN PUMA — Darcy Pattison, illustrated by Kitty Harvill — The True Story of an Orphaned Cub

**SPIDER** — NEFERTITI, THE SPIDERNAUT — Darcy Pattison, illustrated by Valeria Tisnés — The Jumping Spider Who Learned to Hunt in Space

**AMPHIBIAN** — ROSIE THE RIBETER — Darcy Pattison, Nathaniel Gold — The Celebrated Jumping Frog of Calaveras County

**REPTILE** — DIEGO, THE GALÁPAGOS GIANT TORTOISE — Darcy Pattison, illustrated by Amanda Zimmerman — Saving a Species from Extinction

**MARINE MAMMAL** — PELORUS JACK, THE NEW ZEALAND DOLPHIN — Darcy Pattison, illustrated by Eva Dooley — Inspiring a Government to Protect an Individual Animal

**MOLLUSK** — JEREMY, THE ENGLISH GARDEN SNAIL — Darcy Pattison, illustrated by Olga Gonina — Heredity, Citizen Science, and #snaillove

### Starred Reviews
*Wisdom, the Midway Albatross* - Publisher's Weekly
*Diego, the Galapagos Giant Tortoise* - Kirkus

### National Science Teaching Association-Children's Book Council Outstanding Science Trade Books
*Abayomi, the Brazilian Puma*
*Nefertiti, the Spidernaut*
*Jeremy, the English Garden Snail*

### Best STEM Book
*Jeremy, the English Garden Snail*

**MimsHouseBooks.com**

LEARN MORE